The "Queen" Underground

The Era of Naked Mole Rats at Dresden Zoo

Heterocephalus glaber

Christiane Krönke

Zoologischer Garten Dresden

Personal Observations and Experience in Zoo Husbandry

Keeping and Breeding since 1995 – First Successful Breeding 2003

Imprint

This is not a scientific treatise but a collection of practical experience values based on personal observations and many years of hands-on work.

Originally written: 24 May 2014. Revised: 8 March 2024. Book edition: 2 May 2026, English edition 31 May 2026 (translated with AI assistance)

Foreword

Anyone who wants to keep, care for and breed naked mole rats must be prepared to love these little 'naked ones' and should make a daily effort to think their way into those bucket-shovel-equipped heads.

There are animals you love because they are cute. And then there is the naked mole rat.

Mocked as a 'penis with teeth', ridiculed as a 'sabre-toothed sausage', crowned the world's ugliest animal – and yet: anyone who takes the time to look beyond the peculiar exterior of this deeply fascinating species will quickly lose their heart to it. Admittedly, it is a rather curious animal. That is exactly what happened to me. And this book is the result.

Naked mole rats have been kept at Dresden Zoo since 1995. Since I began caring for them, these animals have never let me go. Every birth, every death, every royal power struggle, every tiny baby taking its first bite of solid food – all of it has etched itself deeply into my memory and my heart.

What you will find in this book is not a scientific treatise. It is something more personal: a collection of experiences, observations, mistakes and successes. It is the record of a long friendship between a human being and the naked mole rat groups in my care.

Perhaps it takes a certain degree of obsession to truly want to understand these animals. But the more secrets they reveal, the more one learns to love them in all their very own, quite extraordinary perfection.

Christiane Krönke

Chapter 1: The Species – A Creature from Another World

The naked mole rat, Heterocephalus glaber (RÜPPELL, 1842), belongs to the family Bathyergidae (African mole rats) and is the sole species of its genus. It is a rodent – but a biological rule-breaker that defies almost every expectation one might have of rodents.

Its home is the semi-arid regions of East Africa: the south of Kenya, Somalia and Ethiopia. There it lives in branching underground tunnel systems, never exposed to daylight, never to rain, never to wind. Its life unfolds in a self-chosen darkness, in narrow passageways shared with hundreds of conspecifics.

The naked mole rat's skin is wrinkled, pinkish and almost entirely hairless. Its eyes are tiny and largely non-functional – why would you need eyes if you never live in the light? Its incisors, however, are powerful and omnipresent. It digs with them, eats with them, and fights with them. The lips can be sealed behind the teeth to prevent soil from entering the mouth while digging – a small, elegant marvel of evolution.

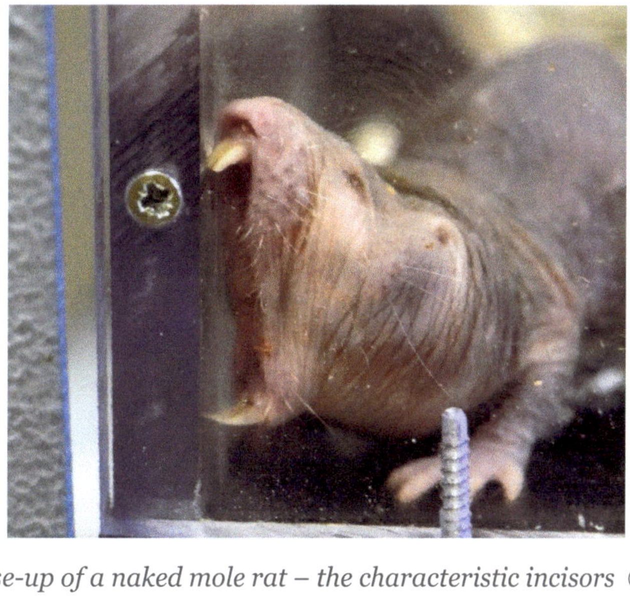

Close-up of a naked mole rat – the characteristic incisors © Ch. Krönke

Physiological Characteristics

Naked mole rats have a four-lobed liver, which gives them exceptional flexibility inside their narrow tunnel systems. Small vibrissae – sensory hairs – are found not only on the face, but also on the tail, hind legs and between the toes of the front feet. With these fine hairs they detect the smallest vibrations in the ground and navigate in complete darkness.

What continues to fascinate biologists: naked mole rats are to some extent poikilothermic – they can adjust their body temperature to the ambient temperature, a capacity unmatched by any other mammal to this degree. They cool down in the cold and warm up in the heat. This is precisely why temperature management is so critical in zoo keeping.

They are also just as agile moving backwards as forwards – an invaluable ability in a narrow tunnel.

Remarkably, female naked mole rats continue to produce eggs into old age – unlike any other known mammal – meaning that even very old alpha females can still reproduce. The documented lifespan of naked mole rats can reach up to 28 years, which is extraordinarily long for an animal of their size.

Naked mole rat alpha females have yet another remarkable trick up their sleeve. With each successive litter, litter size increases – from four to eight pups at first birth to an average of around twenty in later litters. A considerable number of small bodies that must simultaneously fit inside the mother's abdomen – in an animal that moves through narrow passageways every day. Nature has devised an elegant solution: during the first six births, the vertebral bodies of the alpha female actually elongate – she literally grows with her responsibilities. The spine extends so that the belly can accommodate more offspring without compromising the mother's mobility in the tight tunnels.

This change is also visible on the scales: young alpha females weigh on average around 46 grams, while older, experienced alphas weigh around 60 grams on average – a quiet but unambiguous indication of each animal's physical history. As far as is known, this too is unique in the animal kingdom.

A further distinctive feature concerns the number and arrangement of nipples: in naked mole rats, these are often asymmetrically distributed – not mirrored evenly on left and right. The maximum nipple count is 12, and the number can change over an individual's lifetime – something not known in other mammal species.

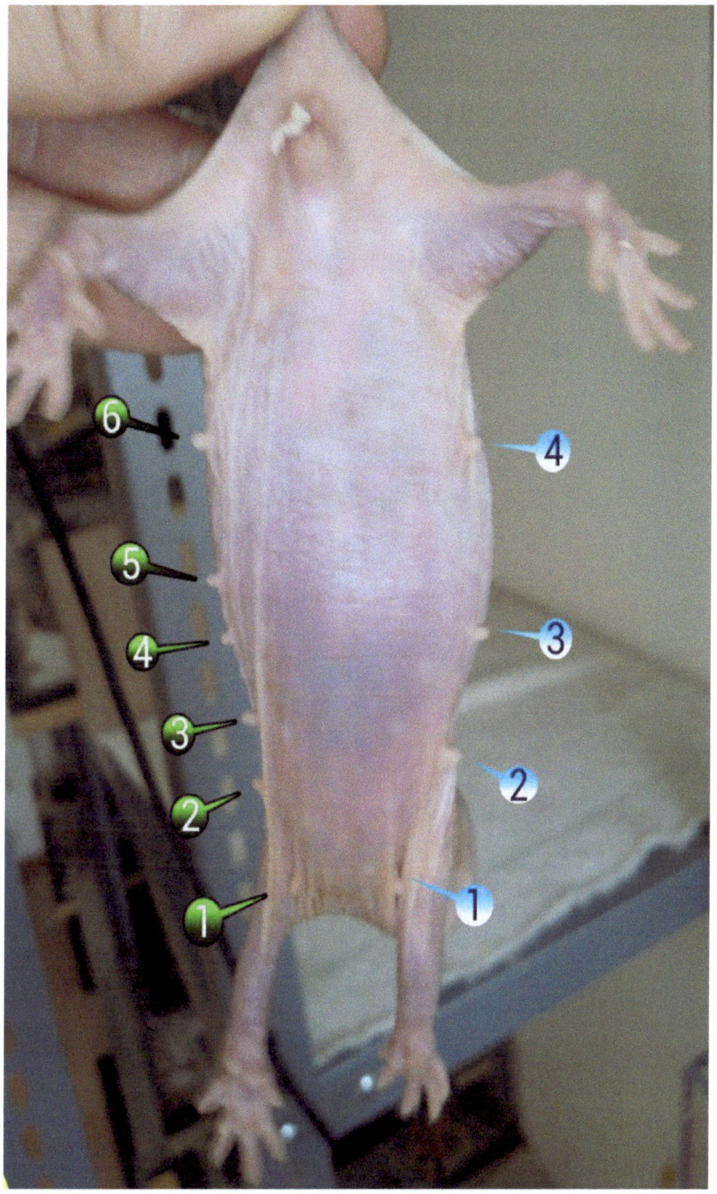

Ventral side showing nipple arrangement – asymmetric distribution © Ch. Krönke

The Social System: A Monarchy Underground

What truly sets the naked mole rat apart is its social structure. It lives in a highly developed community system that is unparalleled in the mammalian world. At the top stands the alpha – the "queen". She alone – with very few exceptions – is capable of bearing young.

All other females are placed under such chronic stress by the alpha's constant dominance behaviour that their reproductive hormones are suppressed. They cannot become pregnant as long as the queen rules, pushes, shoves and controls.

The remaining group members – known as helpers – are assigned different roles according to their size and body weight. Smaller animals take care of food procurement, cleaning and transporting pups. Larger, heavier helpers protect the group – if necessary by physically blocking the tunnels against intruders with their powerful bodies.

Until a few decades ago, these groups were compared to bee or ant colonies, with fixed castes and predetermined roles. But this comparison is flawed. Any female can theoretically become the alpha. The castes are not birth destinations – they are snapshots of a living hierarchy.

Chapter 2: The Beginning – Starting a Group

At the start there is always a group. You either receive one or you assemble one. Sounds simple. It is not.*

The most important rule: a minimum of three animals. One large female, one large male, and one helper – the sex of the helper is irrelevant, but its lower body weight compared to the others is not.

The reason lies in the question of dominance. Equal weight ratios quickly lead to instability in the hierarchy. Two females of the same weight class both vying for the crown is a recipe for war – not a minor scuffle, but a relentless fight in which one of them frequently does not survive.

The future alpha female should weigh approximately 44 to 46 grams. The future alpha male should be between 36 and 46 grams. All other animals in the starting group must be lighter. Only in this way is the hierarchy clear from the outset – and clarity is everything.

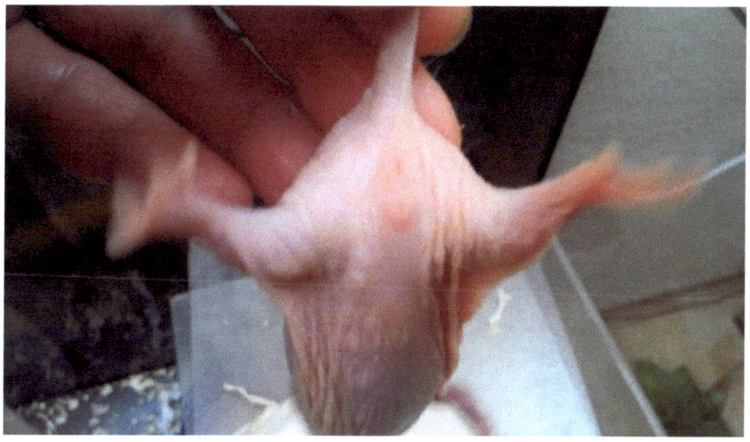

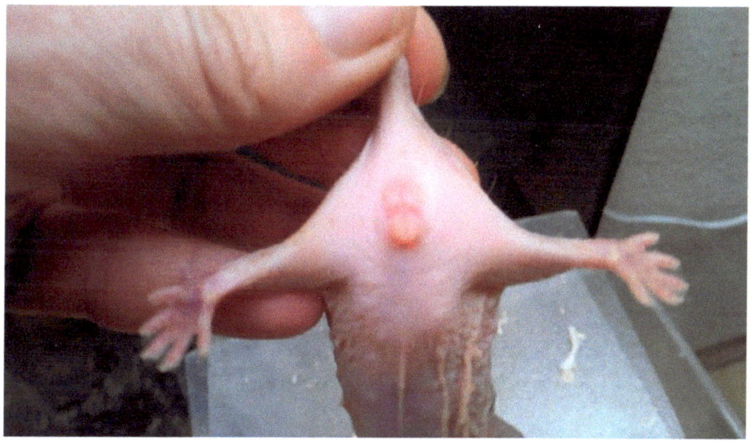

Male (top) and female (bottom) – sex determination © Ch. Krönke

**Literature on group formation from multiple groups: Ludwig, W. & Krönke, C. (2012) Group formation and reproduction in naked mole rats (Heterocephalus glaber) at Dresden Zoo. In "Der Zoologische Garten", Vol. 81, Issues 2–3, 2012, pp. 96–112*

The Enclosure: More Than Just a Cage

A naked mole rat enclosure is not a cage. It is a city. A small, finely structured city under acrylic glass, with bedrooms, toilets, work rooms, kitchens and connecting corridors.

For smaller groups, six to eight chambers are recommended: one or two sleeping boxes, a latrine, a feeding box, and additional rooms for resting or undisturbed eating. As the group grows, the answer is not simply to add more chambers – an enclosure that is too large can no longer be controlled by the alpha without stress, which leads to unrest. Larger chambers are preferable to more chambers.

Differences in height between chambers are valuable. Not only because they imitate natural tunnel systems, but because with a single heat source they create a temperature gradient: warm air rises, and the animals can find the temperature they prefer for themselves.

The sleeping box requires sufficient floor space – for 30 to 35 animals, approximately 18 x 15 cm is adequate. Too small, and pups risk being crushed. Too large, and the group gets cold because body heat cannot be retained.

The Latrine

The size of the latrine chamber is significant: as the group grows, it must handle an increasing volume of urine and faeces. Over time, urine breaks down to produce ammonia, and if ventilation is insufficient, concentrations can build up to toxic levels. Fatalities in other collections have been attributed to exactly this cause. A common misconception is to simply fill the toilet box with large amounts of wood shavings to absorb the moisture – this helps little, however, because naked mole rats rearrange the substrate throughout the enclosure as they see fit, entirely regardless of human ideas about order. It is therefore important to ventilate the latrine box adequately, but not to clean it too thoroughly – a certain amount of the group's own scent must be preserved, otherwise the animals will begin re-marking, causing unnecessary unrest.

Enclosure Design: Height, Ventilation and Structure

Differences in height between the individual boxes and chambers are ideal for a naked mole rat enclosure. It is important for the animals' wellbeing to offer them not just "left/right" movement but also "up/down" – just as in their natural tunnel systems. Additionally, with a single heat source, a temperature gradient automatically develops: warm air rises, and the animals independently seek out whichever chamber suits them best. The warmest box is not always the preferred one – occasionally a cooler "room" is used for a change of air.

A particularly well-conceived ventilation solution came to my attention in a photograph of a European facility: blind tubes in a T-profile had been inserted into the tunnel system, which could be opened and closed as needed. This allows the ventilation of individual chambers to be adjusted specifically and flexibly – an elegant approach I consider well worth emulating.

Humidity and Temperature

Adequate air humidity is essential for naked skin. Because the animals receive wood shavings as substrate in their enclosure, the skin would quickly resemble parchment if it did not receive sufficient moisture. In the natural earth burrows of the African savannah, this is regulated by the soil's own microclimate, which can store a portion of moisture (literature suggests up to 80% relative humidity). In zoo keeping, this must be actively managed!

At the beginning of the Dresden husbandry, moistened substrate was tried, but this quickly caused discomfort for the animals, as wet skin unfortunately also releases body heat. The negative effects were particularly pronounced in juvenile animals, as they lost body energy too quickly and in many cases this led to their deaths.

Over the years, a simple and elegant solution has been established in Dresden: damp sheets of kitchen paper towel are hung in the lids of the chamber boxes. The paper releases moisture slowly into the air without the animals' skin becoming directly wet. This is crucial, because wet or damp skin releases heat, which can be particularly dangerous for juveniles. In this way the naked mole rats' skin remains supple and well-conditioned, without the animals getting cold or drying out. The measured humidity in the enclosures is between 45 and 65 percent – a range that has proven ideal in practice.

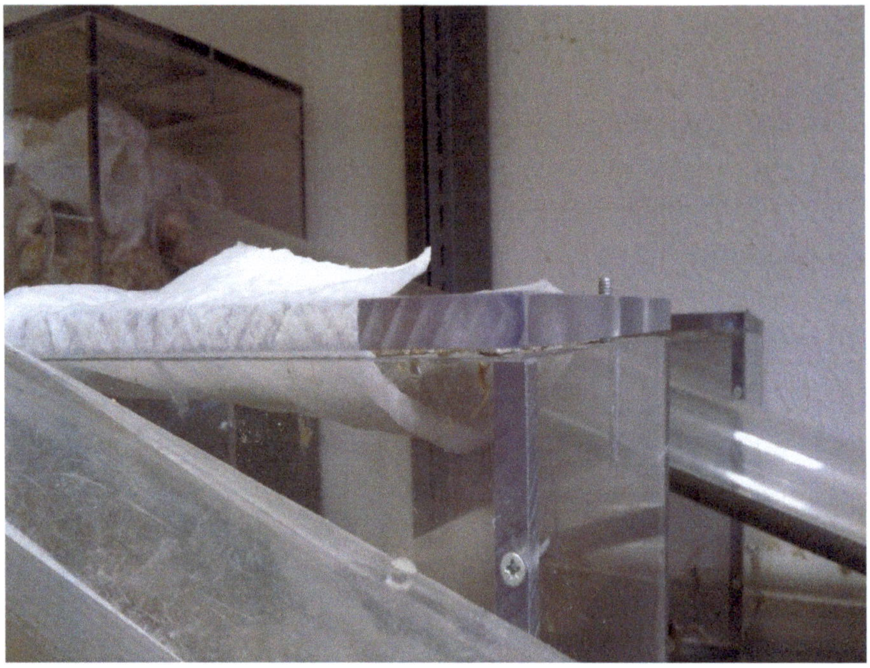

Temperature should be maintained between 26 and 32 degrees Celsius. Sudden temperature drops must never occur. In combination with damp skin, naked mole rats can quickly fall ill.

The most important cornerstones of good husbandry are stability and inner calm within the naked mole rat group. External stress or disturbances are absorbed much better when the group structure is internally stable. Stress within the group – triggered for example by dominance problems or insufficient food – has far wider consequences than one might expect. Pregnant females can lose their pups through chronic stress; it can affect milk production; some individuals may lose weight; pups and juveniles die when persistent unrest prevails in the group. Stress can also be triggered by excessive handling by the keeper. At Dresden, enclosures are cleaned once daily, combined with simultaneous feeding.

To minimise acoustic stress – caused by visitors, work noise or cleaning machines – a radio runs 24 hours a day in the Dresden animal rooms. The animals become accustomed to a steady background and respond less sensitively to brief noise peaks.

The Nest: Teamwork and Toilet Paper

Naked mole rats love building nests. They therefore regularly receive toilet paper – but neither shredded nor placed directly in the sleeping box. The sheets are not only nesting material; they are also enrichment. Working together in well-organised cooperation, the animals transport the sheets collectively into the sleeping chamber. Anyone who has watched a small naked mole rat, with an earnest expression and great concentration, pulling a sheet of toilet paper through a tunnel that is three times its own length, will immediately understand why it is an excellent form of enrichment – and why one simply cannot help loving these animals.

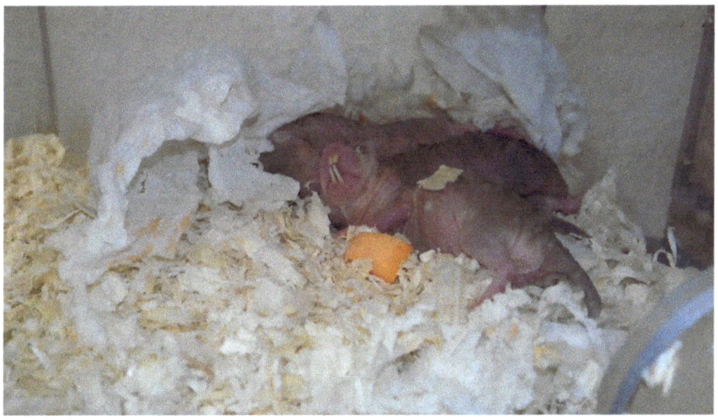

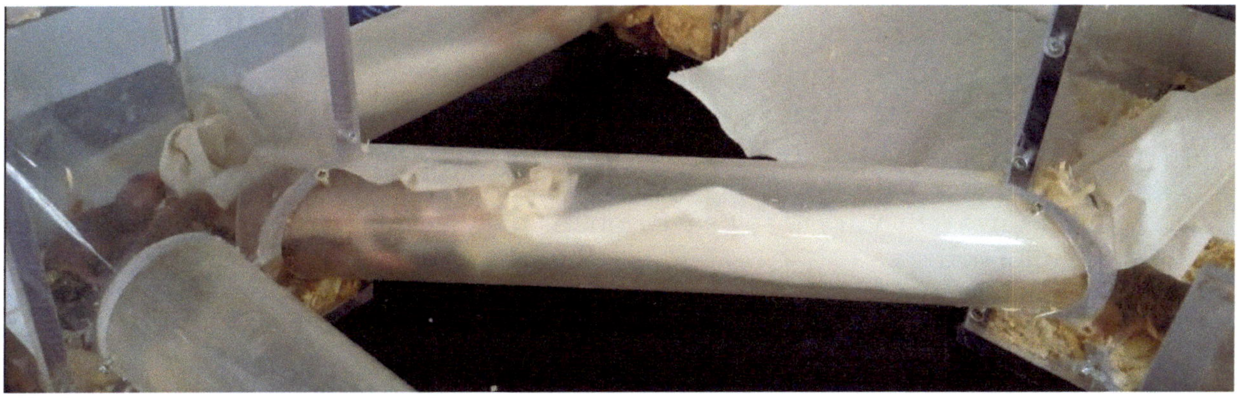

Naked mole rats cooperatively building their nest with toilet paper © Ch. Krönke

Cleaning the Enclosure

Cleaning the enclosures should involve removing old food and damp substrate, wiping moisture from the box walls and cleaning the latrine box. The latrine should not be cleaned too thoroughly, to avoid removing the group's scent – otherwise the animals would have to re-mark, which ultimately leads to unrest again. At Dresden, cleaning is carried out once daily, combined with simultaneous feeding, to keep disturbances to the group to a minimum.

Chapter 3: Feeding – What Naked Mole Rats Eat

Naked mole rats are rodents, and their natural diet consists primarily of fibrous plant tubers which they locate underground in the African semi-desert. This dietary biology determines what is offered in the zoo.

The foundation of the feeding plan at Dresden Zoo consists of vegetables and tubers: carrot, celeriac, kohlrabi, parsnip, turnip, sweet potato and Jerusalem artichoke. A small proportion of fruit (a maximum of one small piece per animal) is included as a vitamin source; carbohydrates and starchy foods such as boiled potato and cooked maize complete the meal and are essential for the animals' healthy activity.

Rat pellets are offered four times a week. These can remain in the enclosures if they are not eaten; as long as they do not absorb moisture and begin to mould, they serve as a useful reserve in case other food has run out.

With a balanced and varied diet as described here, additional vitamin or mineral supplements are not required – the animals meet all their nutritional needs through the food provided.

What Naked Mole Rats Love – and What They Don't Need

Banana, mango, papaya, melon, pear, peach – these are the luxury items on the naked mole rat menu. Their eyes virtually light up – if one can say that of nearly blind animals – when banana or pear appears in the food box. But precisely these favourite foods should be given in moderation. They supplement, but do not replace, the far more important tubers.

Oats are a particular treat but should be given rarely to prevent the animals from becoming overweight. The same applies to banana – it is delicious, but not a main course.

The amount of food is always based on an average derived from observed food intake over the preceding days, and should err slightly on the generous side (approximately 8–10 g per animal). As already mentioned, food shortage creates unrest in the group, as logically all animals begin moving around searching for food. Naked mole rats eat throughout the day according to appetite, so feeding must cover a 24-hour period.

The Calcium Problem

Following ultrasound examinations of our animals conducted by the IZW (Leibniz Institute for Zoo and Wildlife Research, Berlin), increased deposits in the kidneys were identified, prompting a more thorough review of the basic feeding programme. Since naked mole rats do not drink, calcium is not flushed out through kidney function as it is in other mammals, suggesting that the kidney deposits may result from higher calcium intake.

Since this has not yet been sufficiently investigated scientifically, no well-founded hypothesis can be established. However, it appears that by living underground, naked mole rats metabolise the components of their diet differently from other mammals we know. We have moved to reducing foodstuffs with very high calcium content (e.g. broccoli, parsley, fennel, Swiss chard) and increasing the use of calcium antagonists (tannins, pectins, proteins). The aim is to minimise deposits and reduce calcium binding in the body – but not to eliminate it, as it is of course needed for tooth and bone development.

Calcium antagonists help reduce calcium deposits. Tannins found in bark are particularly helpful – where possible, we provide small twigs up to 12 cm long approximately twice a week (beech, maple, fruit wood), which are accepted with genuine enthusiasm and systematically dismantled by the animals. In the wild, it can be assumed that root material is gnawed for similar purposes.

Pectins are found in apples and carrots – apples are given 3–4 times a week, carrot daily in a proportion of up to one quarter of the total food quantity.

Potassium, which counteracts calcium (preventing excessive calcium excretion), is found in various fruits. The fruit varieties already mentioned are offered in rotation with apple.

Protein is provided via the aforementioned rat pellets and cooked chicken meat. The meat (preferably skin and cartilage – no bones!) is offered three times a week and is particularly favoured by animals with elevated energy requirements (thin, pregnant or nursing animals).

Beetroot is only fed cooked. Since it stores significant quantities of nitrates – which can form toxic nitrites if incorrectly stored – and is barely eaten when raw, freshly cooked beetroot is offered seasonally. On the naked mole rat menu, it falls firmly under the category of "delicious".

Average food quantity for 7 animals – annotated overview of foodstuffs © Ch. Krönke

Since, as already mentioned, naked mole rats do not drink, their water and moisture requirements must be met through food. They therefore receive cucumber twice a week in addition to their regular diet. Since cucumber is not among their favourite foods and spoils quickly at high temperatures, it appears on the menu only in limited quantities.

Clay: The Underestimated Superfood

In recent years, clay has established itself as a valuable dietary supplement. Crushed clay bricks are accepted with genuine enthusiasm by naked mole rats. The minerals contained in the clay support the animals' health, and the clay dust benefits their bare skin. It can be assumed that in the wild, similar mineral intake occurs through gnawing on root material.

The Death Trap: Seeds and Cardboard

A warning for all who wish to keep naked mole rats – and one that has grown from painful personal experience: foodstuffs containing small seeds are dangerous. The small kernels are not digested and can in the worst case cause intestinal blockages. Kiwi and strawberry are therefore completely absent from the menu. Tomatoes, peppers and grapes only without seeds. We do, however, distinguish between digestible and non-digestible seeds: apple seeds and cucumber seeds, for example, are chewed up, eaten and well digested. Note: since apple seeds contain small amounts of hydrocyanic acid (cyanide), excessive quantities should be avoided to prevent health problems. In small amounts, however, our animals have tolerated them very well in our experience.

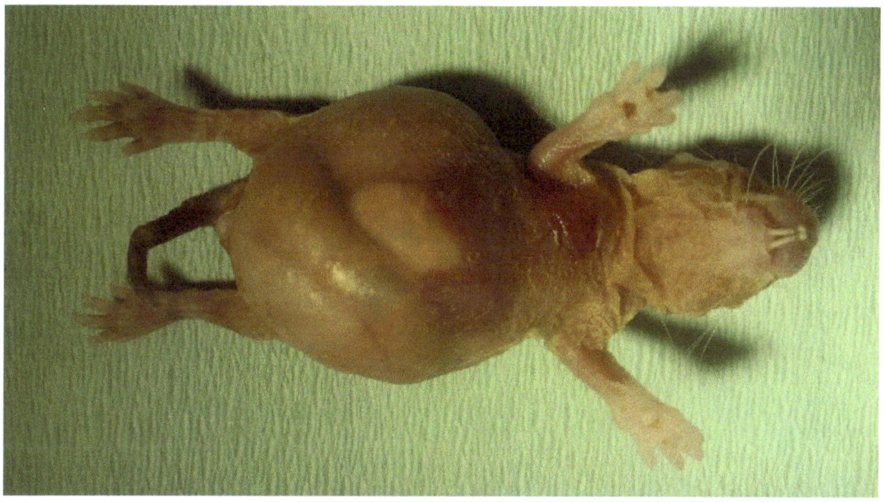

Approximately 11-week-old naked mole rat with suspected intestinal blockage © Ch. Krönke

The same applies to cardboard as enrichment material: the adhesives it contains quickly cause digestive problems. Hard or fruit wood is ideal – soft wood fibres in the bark can also cause problems.

Feeding adjustments for groups with juveniles are addressed in detail in Chapter 5.

Chapter 4: Mating and Pregnancy

For a long time, mating behaviour in naked mole rats in zoo settings was a mystery. When do they mate? How often? Under what circumstances? The questions went unanswered simply because there were no observations.

This has changed. In recent years, mating behaviour was observed and documented for the first time at Dresden during the course of the day. A clear ritual is apparent: the male courts the female persistently; she repeatedly moves into another chamber, then regularly presents herself with her hindquarters raised. The mating act itself is brief but repeated multiple times. A courtship phase, pauses, repetition.

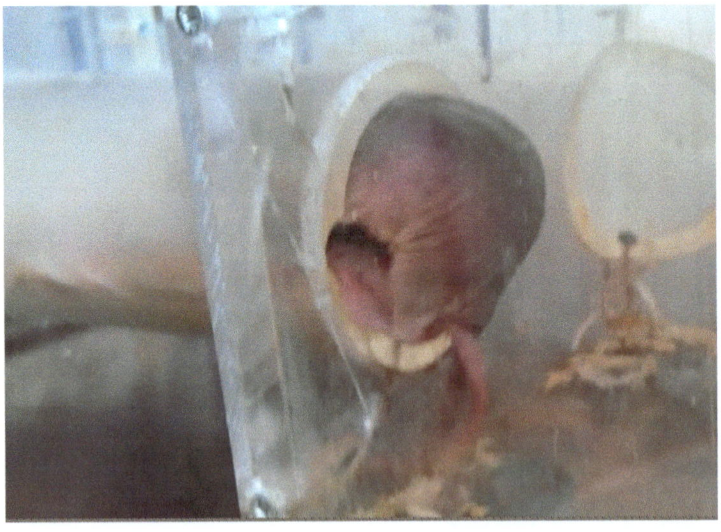

Mating behaviour – documented for the first time during daylight hours © Ch. Krönke

Pregnancy: The Alpha Grows

In a stable group, pregnancy in the alpha female is the almost inevitable consequence. The challenge lies in recognising it early. In the first weeks, this is barely possible visually. The expectant mothers are therefore weighed regularly in Dresden.

A young alpha female gains approximately 50 to 60 percent of her original body weight by the time of birth. Older alpha females who have already had several litters can gain up to 70 percent additional weight – reaching well over 100 grams on the scales. The record weight during an advanced pregnancy was 118 grams.

First litters typically comprise four to eight pups. Later litters can reach up to two dozen pups according to literature. At Dresden, the average is 14 to 18 juveniles. The current record: 28 pups in one litter, of which five were stillborn.

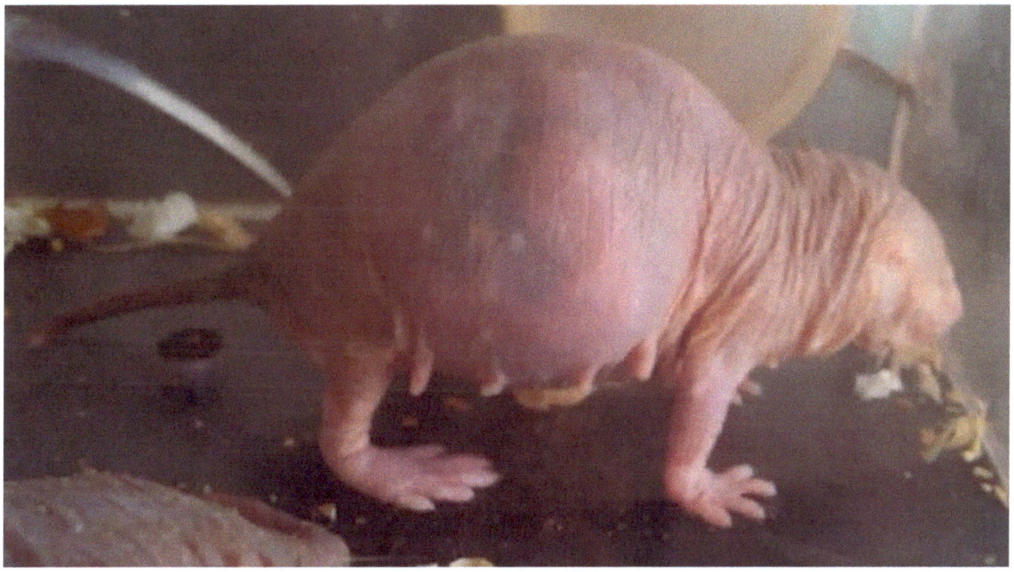

Young alpha female in advanced pregnancy © Ch. Krönke

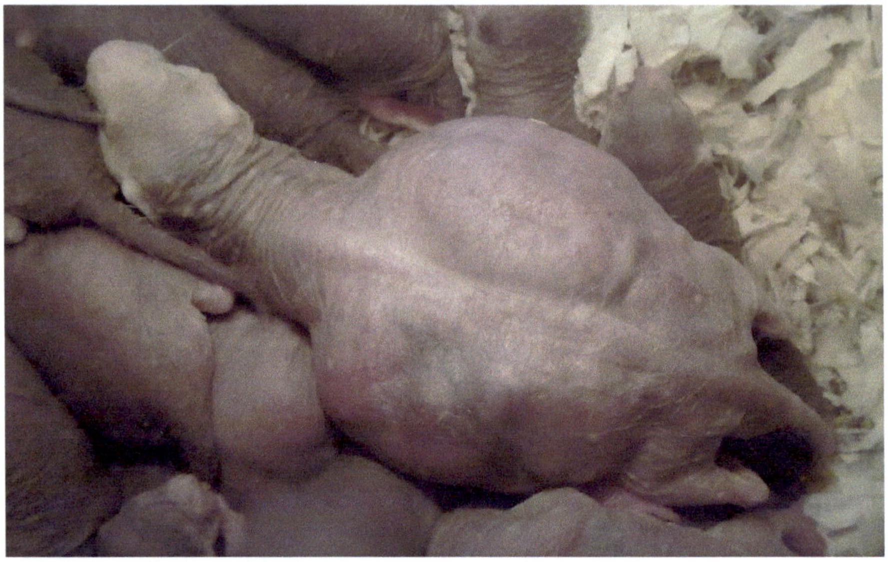

Older alpha female in advanced pregnancy – comparison © Ch. Krönke

Chapter 5: Birth – Early Morning and in Complete Quiet

Births usually happen in the early morning hours. With large litters the process can extend over four to six hours. The mother does not simply lie still during this time – she moves. She walks around the sleeping box, and it can be assumed that this movement facilitates the birth. Pups sometimes arrive tail-first. Other group members immediately help clean the newborns.

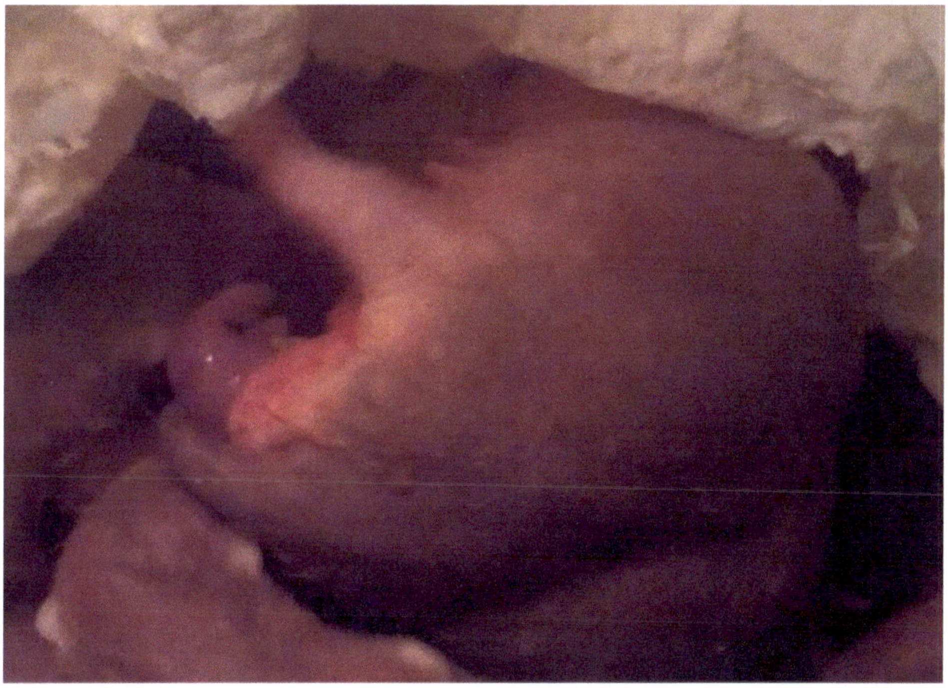

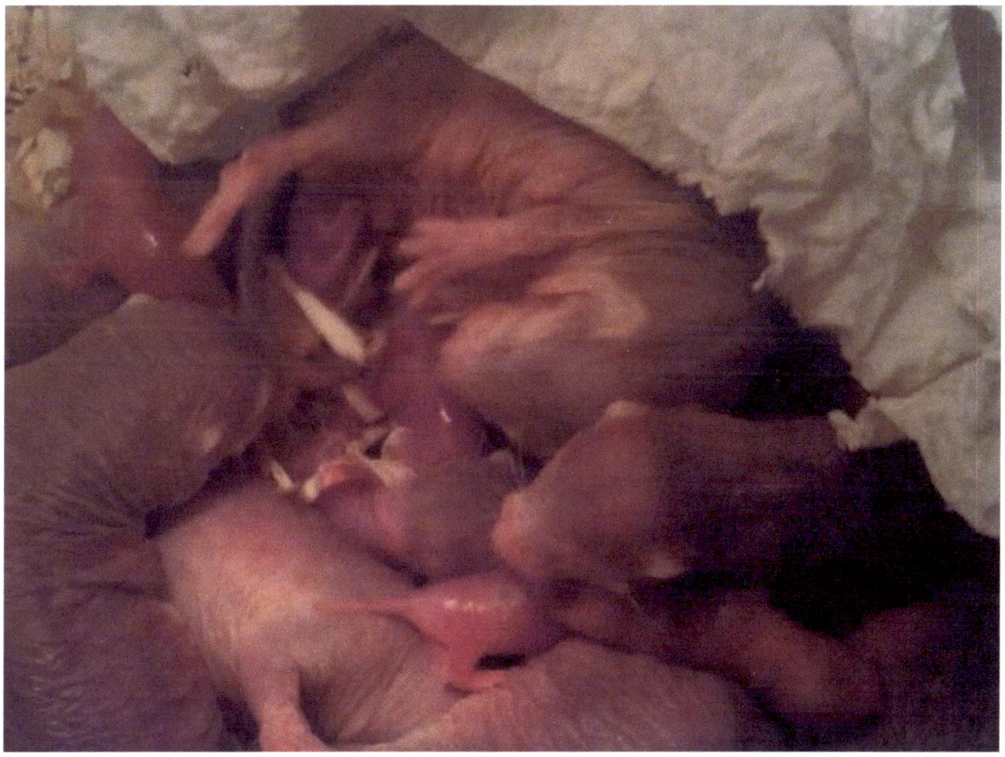

The alpha during birth – other animals clean already-born pups © Ch. Krönke

It is self-evident that the group must remain in complete calm during this time. From a husbandry perspective, disturbances on this day should be kept to a minimum: cleaning work is shifted to late afternoon, and only the essentials are carried out (removing old food remnants from other chambers and roughly cleaning the latrine). Fresh food goes in the feeding chamber and that is all. Quickly perishable items such as peppers, tomatoes or cucumber should be withheld for the time being.

Nevertheless, foodstuffs with higher moisture content are important to allow the alpha to replenish the deficit caused by the birth (e.g. sweet potato, fresh or cooked pumpkin, celeriac, parsnip). When cleaning the enclosure in the following days, the mother and pups should be disturbed as little as possible; if food from the previous day must remain in a box, at least choose items that only dry out (e.g. carrot) rather than becoming soggy. It is particularly important to ensure that sufficient food is provided at all times. If the group becomes unsettled due to food shortage, this can have devastating consequences for the pups, as they have neither peace nor opportunity to nurse. In some cases, food shortage can also lead to the juveniles being eaten.

The First Week: Anxious Observation

Naked mole rat pups can survive for up to seven days without milk. This sounds reassuring. It is not. This first week is the most critical. Everything depends on whether the mother has milk, whether the group is calm enough for the pups to nurse, and whether the sleeping box stays dry.

An unsettled mother – stressed by dominance conflicts, noise or excessive disturbances – may not produce milk. When the milk ridges are not visibly swollen, when pups crawl aimlessly through the nest, when their abdomens remain flat rather than rounded with a milk-filled belly – there is a serious problem.

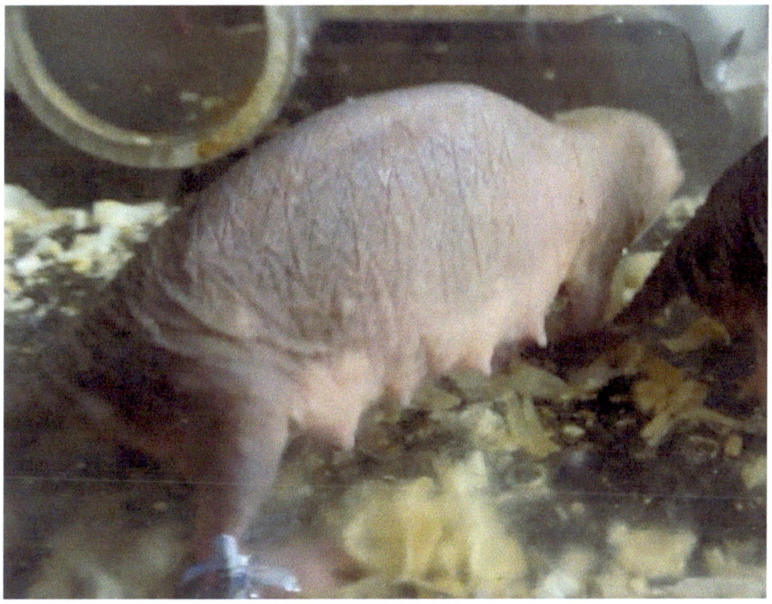

Clearly visible milk ridges of a nursing mother © Ch. Krönke

Waterlogging is another silent killer. Juveniles with damp skin cannot compensate for body heat loss quickly enough. Dry sleeping boxes are not a detail – they are a survival condition.

Newborn naked mole rat pups in the nest – first hours of life © Ch. Krönke

From Milk to Solid Food: Weeks Three to Eighteen

Once the first week is behind them, one can breathe again. In the second week of life, the pups visibly gain body mass. Round milk bellies are clearly recognisable.

Between day 15 and 20, juveniles begin nibbling at food for the first time. They now need light, easily digestible food: boiled potato, banana, pear, mango. Flatulent vegetables such as kohlrabi, radish or spring radish should be avoided initially. In combination with mother's milk, tropical fruits can also cause fermentation.

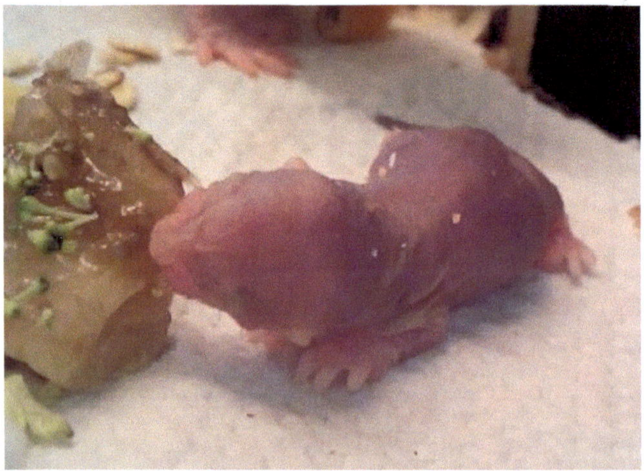

A juvenile aged approximately 19 days eating banana © Ch. Krönke

Cooked carrot is also willingly nibbled © Ch. Krönke

Weight development: at four weeks they have doubled their birth weight of 1.6 to 2.1 grams. At five weeks they weigh around 6 grams, at seven and a half weeks 8 to 10 grams, at twelve weeks 16 grams, at 18 weeks an average of 18 to 20 grams. All of this depends on litter size, group structure and enclosure size.

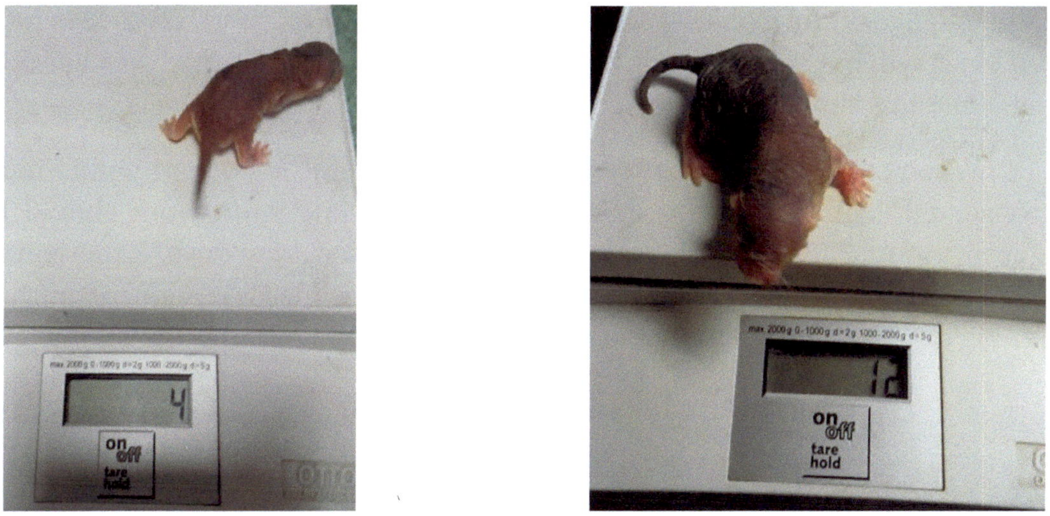

Juveniles being weighed – weight monitoring © Ch. Krönke

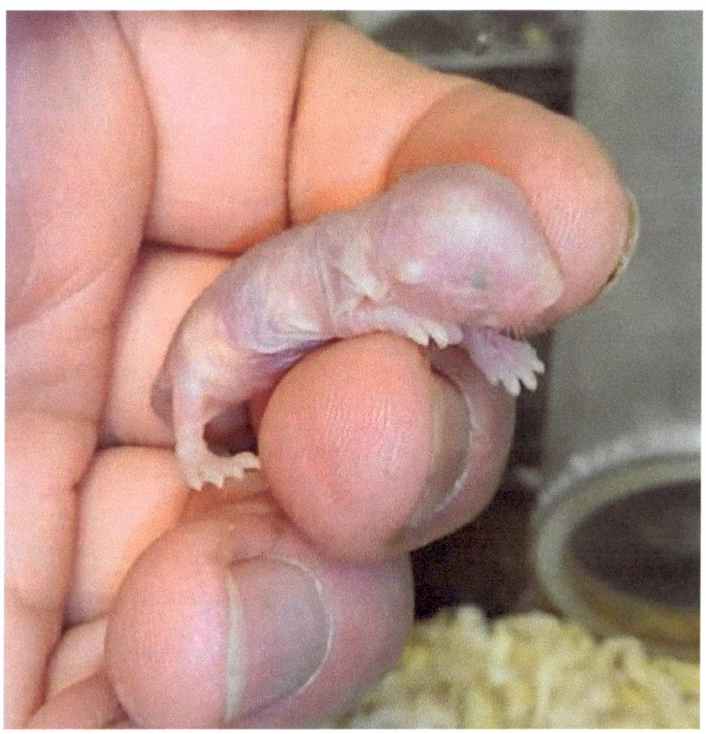

Birth Intervals: An Individual Rhythm

Births occur on average every three months, i.e. approximately every 90 days. However, this is not a rigid pattern – individually, the intervals vary considerably depending on litter size, the number of successfully raised juveniles, and the health and age of the alpha. At Dresden, birth intervals as short as 77 days have been observed, but also intervals of up to 126 days. The alpha thus sets her own pace – and this should be respected.

When pups or juveniles are present, a number of important points must be observed. With young, inexperienced mothers, enclosure work should be reduced to an absolute minimum for at least the first four days. Experienced alpha females are less sensitive – here a slightly reduced version of normal cleaning can already take place the day after birth.

Chapter 6: Temperature Management – A Stress-Free Technique

At some point the sleeping and birthing box also needs cleaning. But how does one get an entire group to relocate without causing chaos?

The solution is as simple as it is clever: heat. Naked mole rats follow warmth. The target box is briefly supplied with extra heat – via a lamp or heat mat – while the current box is left without a heat source. The group follows, relocates, and the box can be cleaned. This can take one to one and a half hours and sometimes does not work at all. But in most cases it is the appropriate and, above all, stress-free solution.

If unrest nevertheless arises in the group that has not settled within an hour of the disturbance, the heat trick works in reverse here too: reducing the temperature by approximately two to three degrees causes all animals to gather in the nest to warm each other – and the pups finally get the peace they need to nurse.

Chapter 7: When Things Go Wrong – Problem Cases from Practice

The naked mole rat is a book with seven seals. Sometimes ten. What follows are real situations from the Dresden collection – without embellishment, because only honest mistakes generate real knowledge.

Case 1: Two Potential Successors

In the established display group, the alpha died along with several other adult animals. The group had to reorganise itself. Two females became pregnant simultaneously and gave birth three weeks apart. Both litters were initially raised. The pups even nursed from the other's mother. It seemed to be working.

Then pups began dying. Both females were pregnant again. The internal stress level caused by the unresolved question of succession was apparently so high that all pups from both litters died. The first female produced no milk ridges with her next litter – the entire litter starved.

The decision: the first female was separated with one large male and four additional small animals. In the new, calm group she recovered. The subsequent litter: 12 of 13 pups survived. The best survival rate in the entire breeding history to that point (before 2014).

Case 2: No Milk

An established alpha female, regular breeding successes – and yet several litters starved. The alpha produced no milk ridges in any case. Repeated brief skirmishes with other animals, never an open war. The suspicion: at least one, probably two, other large females were making life difficult for the alpha. Not through open confrontation, but through constant pressure.

Solution: the alpha female was transferred to a small new group – one large male, one small male, one small female. She recovered. Her next litter was successful.

Case 3: Royal War

Three females, all in the same weight class of 46 to 50 grams, after the death of the old alpha – this is a powder keg. What appears to an inexperienced observer as a lively meeting is in reality systematic bullying. The "rank and file" shift their loyalty to whichever female appears dominant at any given moment.

The lower-ranking animals are loyal to whichever female is currently dominant – one animal (likely the second female) can be seen to the left in the tube © Ch. Krönke

The war was barely preventable. A first casualty. Then the surviving two females in a dramatic fight, both severely injured. Wounds in the jaw and forelimb area are deeper and more serious than the superficial back lesions.

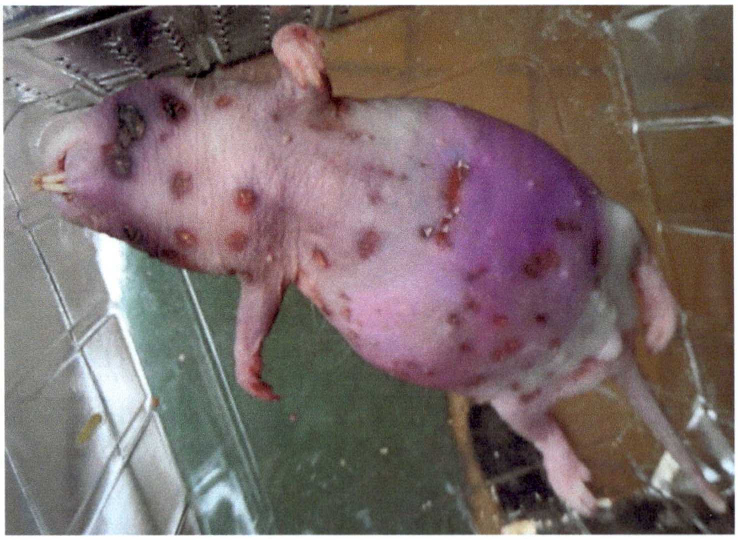

Severely injured female following fights for dominance © Ch. Krönke

Once this point of confrontation has been reached, there is no turning back. Hesitating too long means another casualty.

In the end: the remaining 4 males and 2 females were split into pairs, one male each, forming two new groups – with the hope that at least one of the two groups would be stable enough to successfully raise young. Both groups stabilised and raised their subsequent litters.

Case 4: Pups Dying in a Stable Group

In an established group that had already successfully raised a large litter, pups suddenly began dying. Two causes were identified: the group had chosen a sleeping place directly next to the latrine box – the ammonia concentration may have been too high. Furthermore, the floor area of the sleeping chamber was too small for the number of animals; they were lying in three layers on top of each other, and some pups were presumably crushed.

Solution: the latrine box was better ventilated; the group was redirected via heat manipulation into a larger box. Result: just under 50 percent of the litter could be saved.

Chapter 8: Ventilation and Enclosure Climate – The Invisible Danger

Ventilation is one of those topics that is easily underestimated – which makes it all the more dangerous. When urine decomposes, ammonia is produced. If latrines are not adequately ventilated, concentrations can build up to toxic levels that are life-threatening for the animals. The size of the group must always be kept in mind: the more animals, the more urine, the greater the ventilation requirement.

The hinged lids of the Dresden enclosure chambers contain recesses fitted with mesh grilles for ventilation. Some of these openings have been partially closed with gaffer tape to prevent too much moisture from escaping. The sleeping box lids remain open to prevent waterlogging in the nest. Likewise, the lid openings of the latrine chambers remain unsealed so they can air out properly. It is therefore a constant balancing act: enough ventilation for the latrine, enough moisture for the skin – and dry warmth for the nest.

The display enclosure at Dresden Zoo has been progressively improved over recent years: the floor area of several boxes was doubled and the ventilation openings at the back of the boxes were enlarged – a direct result of successful breeding and the associated growth of the group.

Regarding temperature: the range between 24 and 32 degrees Celsius is well tolerated by naked mole rats, provided no sudden temperature drop occurs. If the animals begin to show signs of feeling cold – recognisable by them pressing unusually closely together and being barely active – and this is combined with damp skin, they can quickly fall ill. Mole rat pups in particular are sensitive to excessively high temperatures. For pups, a range of 26 to 28 degrees Celsius has proven to be ideal.

Chapter 9: Stress Factors – What Disturbs Naked Mole Rats

The most important lesson from years of naked mole rat husbandry is: stability is everything. A stable group can absorb external influences well. An unstable group collapses quickly.

The most important stress factors at a glance:

- Insufficient food: leads to unrest and active searching throughout the entire group.
- Enclosure too large: the alpha can no longer control her territory.
- Dominant females of equal weight: latent or acute risk of war.
- Too many external disturbances: noise, vibrations, excessive keeper interaction.
- Loss of group scent through excessively thorough cleaning of the latrine.

Stress has far wider consequences in naked mole rats than one might expect. Pregnant females can lose their pups. Nursing mothers lose their milk. Pups and juveniles die when persistent unrest prevails within the group. No animal is untouched by what happens in the group.

Chapter 10: Transport – A Group on the Move

At some point the moment comes when naked mole rats must travel. The first two-day overland transport to the United Kingdom was a new challenge for the Dresden collection. What was learned then has since become standard practice.

The transport box must be robust and made of hard plastic. Wooden crates, even when metal-lined, cannot maintain the necessary temperature. For longer transports, the box can be structured with plastic tubes and small cardboard boxes – for transport only, not permanent use, as cardboard is hazardous.

For 30 animals: floor area approximately 30 x 50 cm. The group's own substrate must be included – familiar scents considerably reduce travel stress. Additional toilet paper can be placed in the enclosure the day before departure.

The plastic box is placed inside a larger polystyrene box with a lid. A heat pack (available from reptile supply shops) is placed sideways between the two boxes – this creates a temperature gradient, and the animals can choose where they feel most comfortable. Minimum temperature during the journey: 26 degrees Celsius.Food for one day is included. For longer transports: a second portion for the following day. Quickly perishable foodstuffs such as potato should be omitted. A damp cloth goes on the lid of the plastic box. The polystyrene box must have air holes.

Transport box inside polystyrene packaging – floor area for 30 animals approx. 30×50 cm © Ch. Krönke

Chapter 11: History vs. Today – Brehm's Animal Life and the Modern Perspective

The naked mole rat became known to science comparatively late. In the old Brehm's Tierleben of 1934 – that monumental standard reference work of zoology found in countless German homes – an entry appears on page 230 of the fourth volume that invites a smile today.

Brehm (1934):

"The genus of naked mole rats or bald rats contains the most peculiar little mammals one could possibly imagine – precisely because of their hairlessness."

Today: Nothing much has changed – the naked mole rat still provokes strong reactions. The visual fascination (the 'peculiarity') remains, but our perspective has shifted. While nicknames like 'sabre-toothed sausage' or 'penis with teeth' describe the surface, a closer look reveals one of the most biologically fascinating mammals in the world.

Brehm (1934):

"With their bare skin and their tiny eyes, useless for seeing, they look like blind nestlings of a large animal."

Today: The naked mole rat's eyes are indeed severely reduced – they can perceive light and dark, but little more. What they lack in vision they compensate for with a highly developed system of vibrissae on the tail, hind legs and toes, with which they detect the smallest vibrations in the ground.

Brehm (1934):

"From the not particularly large head the incisor teeth protrude; external ears are absent."

Today: The absence of external ears is simply logical evolution – in a tunnel system they would only be in the way. What is absent on the outside is perfected on the inside.

Brehm (1934):

"The most remarkable feature of the foot are the hair fringes, reminiscent of those on the hind foot of the water shrew. The value of this ciliation, which increases the foot's surface area without increasing load, is obvious for an animal that spends its whole life burrowing in loose sand."

Today: Brehm was remarkably correct here – the hair fringes on the feet do indeed assist with digging in loose substrate. What he could not have known: these vibrissae simultaneously function as sensory organs with which the naked mole rat perceives its entire surroundings in complete darkness.

What Brehm could not have imagined in 1934: this inconspicuous inhabitant of the East African semi-desert would decades later become one of the most intensively researched mammals in the world – for its resistance to cancer, its extraordinary longevity of up to 30 years, its ability to survive in low-oxygen environments, and its unique social structure.

Since 1995, Dresden Zoo has been writing its own chapter in the history of this species. Since 2003, juveniles have been successfully raised. From 2010, the focus deliberately shifted to success-oriented breeding. Today Dresden is regarded as one of the most experienced naked mole rat facilities in Europe.

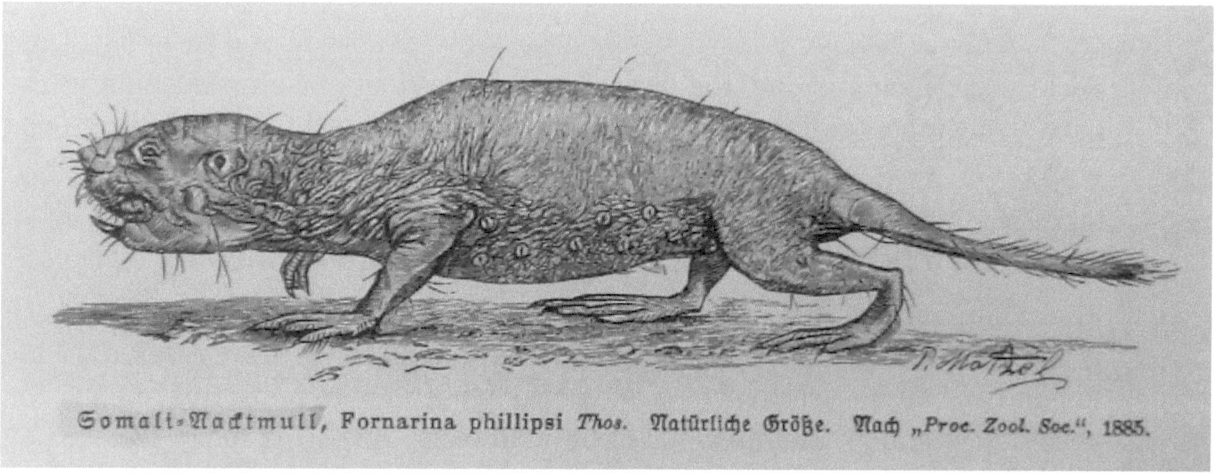

Historical illustration: Somali naked mole rat (Fornarina phillipsi), from Brehm's Tierleben 1885

Addendum: The Social Structure in Detail

The Social Structure of Naked Mole Rats: Between Biological Precision and Living Dynamics

The correct terminology and social structure of naked mole rats is frequently misunderstood. A group of these animals is no longer referred to as a "colony" but rather as a group or community – emphasising the critical difference from insect societies such as those of ants or bees. But the comparison with bee colonies falls short in yet another decisive point: unlike insects, naked mole rats are not born into a caste for life. Every female carries the potential to one day become the alpha. The roles within the group are not a fixed destiny, but snapshots of a living hierarchy.

A System of Helpers and Specialists

The members of a group are today referred to as helpers, with task allocation often correlating with physique. Smaller, lighter animals take on the everyday organisational tasks: they procure food, keep the tunnels clean and transport pups to the nest. The larger and heavier helpers, by contrast, bear responsibility for security. In an emergency – such as a predator attack – they block the entrances to the nest with their powerful bodies. A quiet but impressive sacrifice for the community.

This behaviour is particularly clearly observable in zoo keeping: when enclosure cleaning is due, the larger animals suddenly take up position in the connecting tubes or directly in front of my hand, to protect the rest of the group.

The Role of the Alpha: Dominance and Hormonal Control

The dominant female – the alpha – is the sole reproductive individual in the group, meaning only she is capable of bearing young. In very rare cases, a second female can become pregnant alongside the alpha and raise pups; this remains the absolute exception. The previously common term "queen" is today replaced by "alpha" to better reflect the social dynamics. Together with the alpha pair – which in larger groups can include up to three reproductively active males – she forms the centre of the group.

The mechanism behind her sole reproductive dominance is as effective as it is brutal: the alpha employs a form of 'bossing' to suppress the reproduction of other females. Through constant pushing, controlling and aggressive interactions, she triggers chronic stress in the other females. This stress affects their hormonal processes so massively that their reproductive cycle is stopped. Those who obey the ruler get no offspring. Those who do not face trouble.

The Common Goal: Social Cohesion

What makes the social cohesion so remarkable despite this rigid hierarchy: almost all group members actively participate in nest building, pup rearing and managing the food reserves in the sleeping box. Even younger animals from previous litters integrate themselves early into the group's task structure. Only the task of security remains reserved for the powerful adult animals.

This system is also far more flexible than it first appears. If the alpha is lost or weakened, another female can take on the reproductive role as her hormonal levels normalise. The social system of naked mole rats thus represents an interesting mixture of dominance behaviour, hormonal control and collective care, contributing to their unique way of life.

The Scientific Importance of Zoo Husbandry

Zoos such as Dresden, where naked mole rats have been successfully bred for many years, make an indispensable contribution to research and the conservation of animals that are not easy to keep in zoological institutions due to their specialised lifestyle and needs. Through daily observation and breeding in such facilities, they allow us a deeper engagement with the biological and medical characteristics that are difficult to study in the wild.

Zoos function here as important research centres and breeding stations that continually expand our knowledge of this fascinating social flexibility. It is very likely that the breeding of naked mole rats at Dresden Zoo can deepen insights into social behaviour still further.

A Final Word on Husbandry Competence

According to the **guidelines of David Wood (Mill Mountain Zoo, Roanoke, VA) and Raymond Mendez (Work as Play, Portal, AZ)** in the Husbandry Standards for Keeping Naked Mole Rats in Captivity, keepers who care for the husbandry and breeding of naked mole rats should be experienced in handling small mammals and sufficiently informed about the unique social structure of these animals and the associated care requirements. I can only fully agree. Naked mole rats are not animals to be kept as a side hobby. They demand attention, sensitivity, patience – and yes, a certain degree of obsession. Those prepared to bring all of this will be rewarded with one of the most fascinating insights that the mammalian world has to offer.

Afterword: A Book with Seven Seals

Yes, naked mole rats are a book with seven seals – but intensive observation makes it easier for us to improve husbandry and maintain the health of the animals. Perhaps it takes a certain degree of obsession to want to understand the little naked ones, but the more secrets they reveal to us, the more one can love these animals in all their very own perfection.

Not all husbandry conditions are yet optimal. But the successes of recent years show: we are well on the way.

Personally, I am completely devoted to these little creatures and am always delighted when I can share my enthusiasm with others – colleagues, zoo visitors, animal sponsors.

Acknowledgements

My thanks go to Dr Wolfgang Ludwig, who supported me in my work – surely also with a smile at my ‘obsession’ – and helped me find my way with advice and practical support.

I also thank Dr Dimitri Widmer and Eva Ziemssen, who have stood and continue to stand by me in every veterinary emergency, and Dr Widmer in particular for ultimately giving me the courage to put my observations in writing.

I would like to thank my immediate colleagues in the Africa House, Ronny Moche and David Lück, who keep my back free for all the naked mole rat minutiae and repeatedly help me untangle my occasional mental knots.

My thanks also go to all colleagues in the Africa House who competently look after the naked mole rats in my absence – with special mention of Miriam Grünwald, who follows my deeper interest in this extraordinary species with great enthusiasm.

Thank you to everyone for sharing the interest in our 'sabre-toothed sausages' and for your patience when I had once again dived deep into the world of tunnels.

Christiane Krönke

Zoologischer Garten Dresden

May 2026

About the Author:

Born in 1970, she comes from a true animal keeper family: both parents worked as animal keepers, and her father devoted more than four decades of passionate dedication to elephant care at Dresden Zoo. Her own path was therefore hardly surprising: from 1986 to 1988 she completed her training as a zoo animal keeper at Dresden Zoo — the beginning of a career driven by deep conviction. Caring for animals is, for her, not a profession but a calling. In 2002 she discovered her very own niche: the world of small mammals. Since then, her particular passion has been devoted to this often underestimated group of animals, which she approaches with expertise and dedication.

Imprint

Bibliographic Information of the German National Library: The German National Library lists this publication in the German National Bibliography; detailed bibliographic data are available on the internet at dnb.dnb.de.

Publisher: BoD · Books on Demand GmbH, Überseering 33, 22297 Hamburg, bod@bod.de
Printed by: Libri Plureos GmbH, Friedensallee 273, 22763 Hamburg

ISBN: 978-3-6963-6829-6